BRUMBACK LIBRARY
The food chain /
Penny, Malcolm.   591.5 PEN
JNF   c1

3 3045 00019 9394

j591.5   8902729
PEN
                            11.90
Penny, Malcolm
  The food chain

CHILDREN'S LIBRARY
THE BRUMBACK LIBRARY
OF VAN WERT COUNTY
VAN WERT, OHIO

The Animal Kingdom

THE FOOD CHAIN

Malcolm Penny

Illustrated by John Yates

The Bookwright Press
New York · 1988

The Animal Kingdom

Animal Camouflage
Animal Evolution
Animal Homes
Animal Migration
Animal Movement
Animals and Their Young
The Food Chain

First published in the
United States in 1988 by
The Bookwright Press
387 Park Avenue South
New York, NY 10016

First published in 1987 by
Wayland (Publishers) Ltd
61 Western Road, Hove
East Sussex BN3 1JD, England

© Copyright 1987 Wayland (Publishers) Ltd

ISBN 0-531-18167-7

Library of Congress Catalog Card Number: 87-71474

Typeset by DP Press, Sevenoaks, Kent, England
Printed by Casterman, SA, Belgium

Contents

What is a food chain?	4
The first link in the chain	6
The second link in the chain	8
The third link in the chain	10
When is a chain a pyramid?	12
A food chain in the backyard	14
A food chain in a lake	16
A food chain between the tides	18
A food chain on the Arctic tundra	20
A food chain on the African plains	22
A food chain on the coral reef	24
A food chain in a frozen ocean	26
A food chain in the rain forest	28
Glossary	30
Further information	30
Index	32

What is a food chain?

In the Costa Rican rain forest, a male resplendent quetzal carries a beetle in his beak. He is bringing the insect to the nest, where his chicks are waiting to be fed.

Imagine lettuce growing in a garden. It draws water and nutrients from the soil, and uses the energy from sunlight falling on its leaves to make the food it needs, in order to grow.

A slug appears one night and begins to eat the lettuce. The energy contained in the lettuce becomes part of the slug. Later, a ground beetle scurries up and eats the slug. Now the food from the lettuce that has been absorbed by the slug has passed into the beetle, to give it strength to hunt another slug.

After a time, a shrew corners the beetle and eats it. So the energy from the lettuce is passed down through the slug and the beetle into the shrew. This gives the shrew strength as it bustles about in its non-stop search for something to eat.

One night, the shrew is less cautious than usual and its movements are heard by an owl, which glides silently down to catch the shrew and eat it. The food that the lettuce made from the energy of the sun has passed down the chain and is now keeping an owl warm and well-fed, as it prepares a nest in which to rear its babies.

This sequence of events is called a "food chain." It is called a "chain" because food, or energy, is passed from one creature to another until it becomes part of an animal like the owl, which has no natural predators. The owl is said to be at the top of this particular food chain.

In this book, we shall find out how food chains work by examining the different links in the chain. We shall look at many different habitats, from a backyard to a rain forest. We shall discover that some animals eat plants and others eat flesh, but all depend on one another.

The first link in the chain

*Plants are the vital link in every food chain and provide nourishment for many animals. Australian eucalyptus leaves give energy to koala bears (**below left**). The nectar from some flowers of the Australian rain forest forms the diet of the honey possum (**below right**).*

The most important process in the world, and the basis of all food chains, is photosynthesis. Without it, there would be no food to eat, and no oxygen to breathe. Photosynthesis means "making things from light."

When sunlight falls on the leaves of a plant, its energy is captured by chlorophyll, a green chemical in the plant. The energy is used to join together carbon dioxide from the air and water from the soil to make sugar. When the sugar has been made, oxygen is left over. The plant lets the oxygen escape into the air. If it is a water plant, the oxygen appears as bubbles, which float to the surface.

Most green leaves do not taste sweet, even though they make sugar. This is partly because the sugar is used up at once as the plant grows. However, all the energy is not lost: some of it is built into the body of the plant, and is available to any creature that eats the plant.

Green plants are called primary producers, because they are the first link in the food chain and they produce food. The vast grasslands of the world, and temperate and tropical rain forests are important areas of green plants.

At the other end of the scale, many of the plants are very small. In fresh water, or in the ocean, there are millions of floating algae, some of them consisting of only one tiny cell. To see them, the water has to be filtered (finely strained). When all the water has gone, there will be a faint green smear left behind, made up of some of the most important plants in the world, the primary producers.

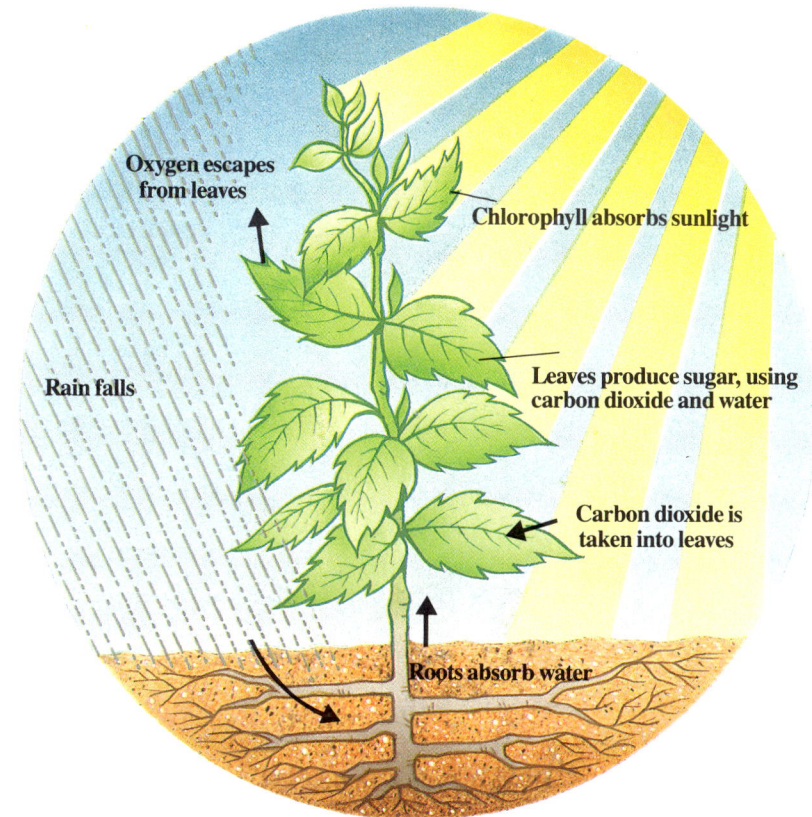

Left *The diagram shows how green plants produce sugar and oxygen by the process of photosynthesis.*

The second link in the chain

The next link in the food chain is made up of the primary consumers. They are herbivores, which means animals that eat plants. On land, these animals can be any size, because land plants can be quite small or very large.

Large plants are often eaten by very small animals, such as the leaf miners and tiny caterpillars that live on oak trees. They may also be eaten by large animals, such as elephants or black rhinoceroses.

Although a grass stalk seems quite small, grass is a very important producer because it grows in such large areas. On the American prairies or the African savannah, grass is the main food of huge numbers of herbivorous animals. Once, the American prairies were grazed by large herds of bison, but now their place has been taken by domestic cattle on ranches. On the African grasslands, huge herds of wild animals still exist, among them wildebeest and many different kinds of antelope.

In water, most plants are very small, like the tiny algae mentioned in the last chapter. However, they are as important as grass is on land, because they feed an enormous variety of animals.

The primary consumers in water are usually very small. They filter the water to extract the plants from it, though some of them are so small that they eat the plant cells one at a time.

Some quite large animals can take a short cut by filtering tiny plants from the water. A good example is the flamingo. Its bill contains special bristles to catch the plants as it scoops up water. An even bigger water-dwelling herbivore is the manatee, which lives off the coast of Florida and around the Gulf of Mexico. It uses its snout to pick clumps of water hyacinth on which it feeds.

Herds of bison graze in Custer State Park, South Dakota. They are an example of a large animal that survives on a diet of grass.

Bristles used for filtering food

Close up of bill

Above *Caterpillars belonging to the family of white butterflies are often found feeding on cabbage plants.*

Left *The flamingo's hooked bill is specially adapted so that the bird can filter the tiny plants in water. The inset diagram shows the bill in greater detail.*

9

The third link in the chain

You will probably have guessed by now that the third link in the food chain is made up of the animals that eat herbivorous animals. They are called carnivores, which means flesh-eaters. We could also call them secondary consumers.

In water they have a difficult job because most primary consumers are so small. A carnivore must be able to find and eat enough tiny herbivores to keep it alive, without wasting energy in hunting for them. This means that secondary consumers in water must be very small themselves. In the ocean the millions of larvae of crabs, lobsters, worms and fish are called plankton. These tiny creatures all feed on even smaller herbivores. Because they are so small, the larvae only have to catch a few small primary consumers to keep themselves alive.

The carnivorous green tiger beetle eats many insects, including garden ants, which are scavengers.

This photograph shows some of the tiny animals found in zooplankton. These feed on even smaller plants floating in the plankton.

On land, the secondary consumers are not necessarily so small. In the soil there are mites and springtails that eat even smaller animals. There are also beetles and small centipedes that can find enough to eat by seeking out large swarms of the smallest herbivores. On a different scale, lions on the African plains do the same when they hunt among herds of antelope.

No consumer can take all of the energy that its food, whether plant or animal, has collected. This is because every plant or animal uses some energy in order to grow. Energy is also lost when animals breathe and when their food passes out of the body.

As we move higher up the food chain, we shall find that this lost, or spent, energy has an important effect on animal numbers, and on how big animals can grow.

When is a chain a pyramid?

As we move up the food chain, we find that the predators get bigger, but fewer in number. At the top of the chain, the really big predators, like sharks or tigers, are quite rare. For this reason, people often speak of the "food pyramid" or "energy pyramid."

At each level of the pyramid, the animals use up some of their energy to live and reproduce, and store the rest in their bodies. This stored energy is what forms the food of the bigger predators. The bigger predators use up some of their energy, so that there is less left for the next level of the pyramid, and so on. Where there is less energy, fewer animals can live on it. These few animals must also be bigger than their prey, or they would not be able to kill it.

This grizzly bear has caught a salmon in an Alaskan river. Gulls are waiting nearby to scavenge any remains left by the bear.

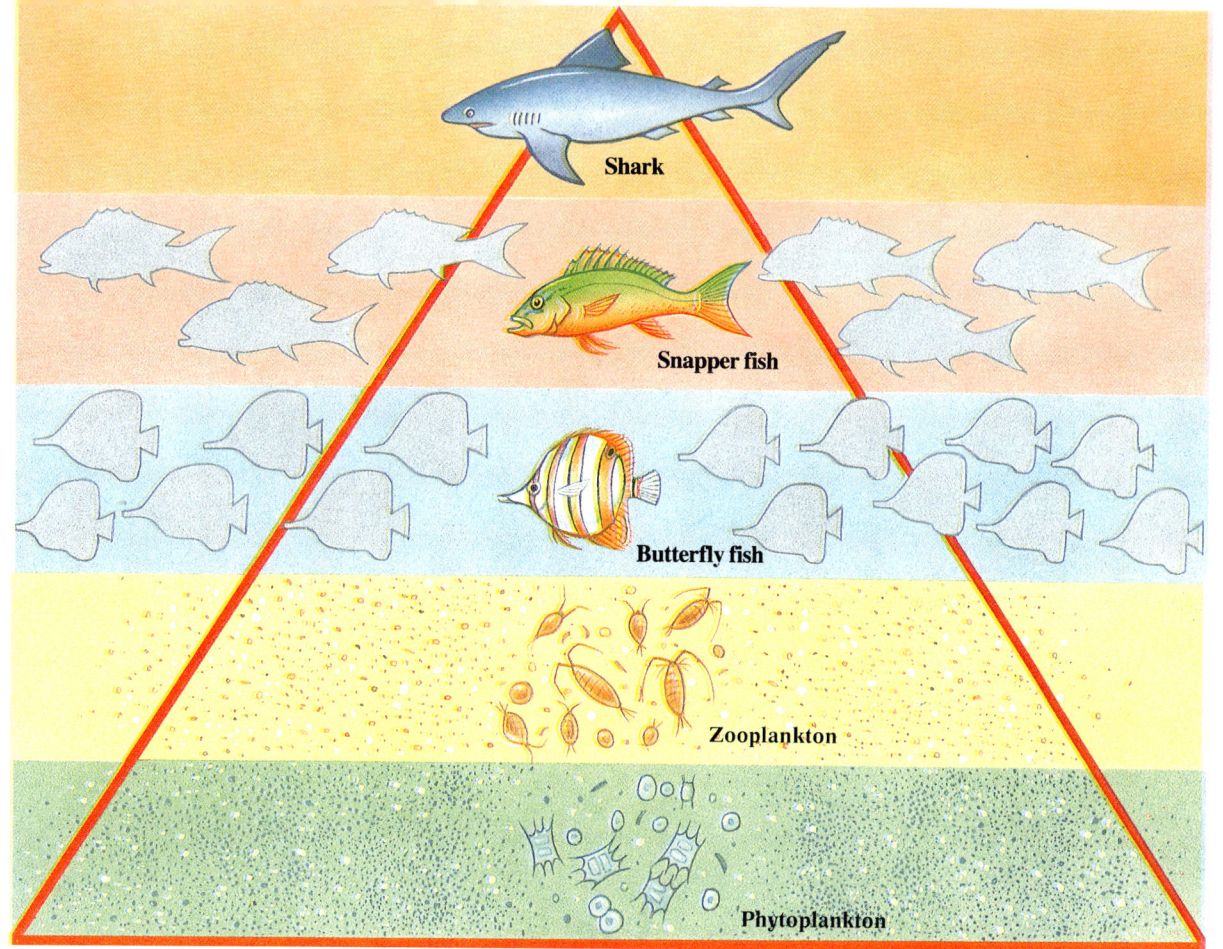

Above *An example of an energy pyramid in a coral reef. Billions of tiny plants (phytoplankton) are eaten by millions of tiny animals (zooplankton). These are eaten by thousands of butterfly fish, which are eaten by hundreds of larger snapper fish. At the top of the chain is a huge shark, which preys on the snapper fish.*

At the very top of the pyramid are huge animals, such as sharks over 10m (32ft) long, which nothing except humans can kill. Any other predator that could catch and kill them would have to be very fast-moving and gigantic. Such an animal would not have enough energy to keep its huge body moving.

The very biggest sharks and tigers are safe from being eaten alive. When they die, of illness or old age, they are eaten by scavengers, such as crabs or beetles, vultures and hunting dogs, or even minute bacteria.

Some top predators are in danger because of the chemicals used to control pests. The chemicals are not used up, but stored in the bodies of animals at all levels of the food chain. At the very top, predators, such as eagles, can receive a poisonous dose accumulated in the bodies of the animals all the way down the chain.

A food chain in the backyard

Now that we know how food chains are made up, we can begin to look at some examples from all over the world, starting with a backyard.

The primary producers are the plants that grow in the yard and garden, from weeds to fruit trees. The primary consumer should be the gardener, but this is not always the case.

The most common plant in most yards is grass, but it is eaten by very few yard and garden animals. Many of the other plants are eaten by a wide variety of primary consumers, most of which we call "pests." Aphids attack roses, fruit trees and beans. Some caterpillars eat cabbages, though others eat weeds such as nettles. Slugs eat almost everything, but gooseberry sawfly larvae eat only the leaves of gooseberry bushes. The list of pests is as long as the list of plants.

Fortunately for the gardener, the secondary consumers are also very active. Both adult ladybugs and their larvae eat aphids, while birds eat caterpillars in large numbers. Shrews and thrushes eat slugs and snails, and so do some ground beetles. Blackbirds hunt for beetles and insects, or may pull worms out of the ground. There are many other links — mice and shrews, which eat seeds and insects, may be caught by the family cat.

At the top of the food chain in a country garden or yard might be a bird of prey, such as a kestrel or a sparrowhawk. In a town, the top predator might be a magpie or a jay, which catch young birds as they leave the nest.

Opposite *In this garden the primary producers are grass and leaves. The primary consumers are the mouse, snail and aphids. The cat, hedgehog, blackbird, ladybug, jay and sparrows are all secondary consumers.*

A food chain in a lake

The primary producers in a lake may be single-celled algae, or stringy pondweed, or larger plants such as reeds and water lilies growing in the shallows around the edge.

The primary consumers may be very small, like the water fleas that eat the floating algae, or they may be slightly larger, like snails or tadpoles that can graze on larger algae. Ducks and even larger animals may be primary consumers, for example, beavers in North America, or coypu in South America.

As the floating plants and the smallest animals die, they sink to the bottom. Their remains may be filtered out of the mud by worms or freshwater clams, or caught on the way down by small fish or water fleas.

Water fleas are very important in a lake. They are near the bottom of the pyramid, and nearly everything eats them, from water scorpions to fish. Dragonfly and mayfly larvae and water beetles find them too small to hunt; they catch tadpoles and small fish, which have fed on water fleas.

When tadpoles grow into frogs, herons come to eat them. When the mayfly larvae emerge, many other birds such as sparrows, wagtails and blackbirds feast on them.

Little fish are eaten by bigger fish, until at the top of the lake food chain is a large fish like a pike. In the southern United States the pike may be hunted by alligators. In North America, big fish may be caught by an otter or a bear. In Australia a large freshwater predator of lakes and rivers is the duckbilled platypus, which feeds on worms and freshwater prawns. So we can see that the top of the food chain often extends beyond the lake.

This food chain could be found in a European or North American lake.

17

A food chain between the tides

In rock pools on the beach at low tide, the animals have many problems apart from finding food. The water may become very warm and there may be no escape from predators. The animals that live between the tides have to be specialized.

However, their food chain works like any other. The primary producers are either small algae, or the large algae called seaweeds. The small algae grow in a thin layer on rocks, where they are grazed on by limpets and sea urchins. The larger algae are browsed on by sea snails.

In rock pools there are many animals that get their food by filtering the water to pick out small particles, the broken down remains of dead animals or plants. This sort of food is called detritus. Barnacles, mussels and fan worms are all detritus-feeders.

Avocet

Lug worm

Mussels

Fan worms

Filter feeders

Crab feeds on sea urchin

The secondary consumers include some unexpected hunters. Whelks bore through snail shells to eat the occupants, and crabs eat sea urchins by reaching in through their mouths to pull out their innards. Sea anemones catch small fish and shrimps in their tentacles.

Crabs are also scavengers, along with small shrimp-like animals called amphipods. Amphipods are very common in muddy places, sifting the mud for food. Lugworms and ragworms, too, burrow in mud and silt.

Like a lake, the rock pool has a food chain that stretches beyond its edges. Oystercatchers crack open mussels, and egrets catch small fish. Wading birds such as redshank and avocets, curlews and ibis use their long bills to dig for worms and amphipods in muddy pools.

Life becomes easier when the tide comes back in again, bringing cool water, and plenty of fresh detritus for the filter-feeders.

The picture shows a food chain within a rock pool at low tide. The avocet catches a worm, which feeds on detritus, while a crab eats a sea urchin. The sea anemone traps a shrimp and the whelk gnaws through the shell of a periwinkle. Barnacles, mussels and limpets cling to the rocks, feeding on detritus.

A food chain on the Arctic tundra

The tundra is the Arctic plain north of the last trees. For most of the year it is frozen solid, but in summer the top layer thaws. It becomes covered with grass, flowering plants and tiny dwarf shrubs, all of which are primary producers. For a few weeks the tundra is alive with enormous numbers of animals.

The primary consumers are insects and grazing animals such as caribou and geese. The young of the geese eat insects during the early part of their life, but many of the insects are blood-sucking flies, which feed on the geese.

The plant shoots are also eaten by voles and lemmings, which breed in very large numbers in some years, but are rare in others. They are the prey of hawks, and especially snowy owls. In a good lemming year, the birds can raise many young, but in a bad year, they go hungry.

Arctic foxes are the main predators on the millions of birds that migrate to the tundra every summer to breed and feed on the many insects.

Caribou, and their cousins the reindeer, move north in spring to feed on the tundra plants. Caribou are hunted by wolves and bears, and by people as they migrate south again in autumn.

The tundra food chain is short, both in time and in its number of links. If one of these links fails, this causes hardship to others. For example, in a year when there is only a small population of lemmings, those animals that feed on the lemmings also suffer. Because it is easily upset, the tundra food chain is often said to be "unstable."

Opposite *Reindeer graze as a snowy owl swoops down on lemmings and an Arctic fox stalks an Arctic tern.*

A food chain on the African plains

Some food chains are stable, because they have so many links that the failure of one or two of them does not cause any problems to the others. A very complicated food chain is found on the Serengeti Plain in East Africa.

The primary producer is thousands of square miles of grass. The primary consumers are insects, such as grasshoppers, and huge herds of grazing animals, such as antelope, wildebeest and zebra. These herds of grazing animals provide food for various predators, such as lions, leopards, cheetahs, hunting dogs and hyenas.

So many animals live on the plains that there are always plenty of dead bodies. The remains are cleared away by scavengers, such as vultures, jackals, burying beetles and fly larvae.

The grazing animals produce large quantities of dung, which beetles use to feed their young.

A hyena feeds on a dead antelope, as scavenging vultures look on. Hyenas usually hunt their prey in packs, but also scavenge off carcases they find. In the background a black rhinoceros browses on acacia branches.

The honey badger specializes in digging up the beetle grubs when they hatch. Storks and egrets are among the birds that eat the adult beetles when they emerge from their underground burrows.

There are some very big primary consumers, such as rhinoceroses and elephants. Like the elephant, the black rhinoceros browses on shrubs, whereas the white rhinoceros grazes on the open grasslands. Such large, heavy animals can only be herbivores: they could not live on food that they had to chase. Because they are so big, they have no enemies to run away from – except humans.

Another important loop in this complicated chain contains seed-eating animals, such as birds and mice, which are eaten by snakes and hawks.

In such a large area, where the animals can wander to find food, there are many alternative links that can be used if others fail.

In Africa herds of springbok graze on the lush grass that grows in the rainy season. The photograph was taken in Etosha National Park, Namibia.

A food chain on the coral reef

Coral grows in clear, shallow seawater that is warmer than 21°C (70°F). The coral animal is like a very small sea anemone. As it grows, it builds a limestone tube round its body. Millions of these tubes, growing in colonies, form a coral reef. The reef has a food chain as complicated as that of the Serengeti Plain in Africa, with one very important difference: the primary producers include animals.

Most species of coral animal have plant cells living within their bodies. They form a partnership, in which the plant cell takes salts and carbon dioxide from the animal, and uses sunlight to produce food and oxygen, which the animal uses.

Other primary producers are algae that grow on the coral. Snails and sea urchins graze on the algae, and are hunted by fish, crabs and starfish.

Shark (top predator)

Cerise grouper fish (feeds on smaller fish)

Moorish idol

Cowrie

Powder blue surgeon fish

Queen trigger fish (feeds on crabs and shrimps)

Top shells

Parrot fish eat coral, crunching it up with a bony mouth, which looks very much like a parrot's beak. Their droppings contain ground-up limestone, which falls to the bottom as silt. In the silt layer there is a new food chain, including burrowing snails and the flatfish that feed on them, and fan worms which live on detritus.

Detritus and plankton are as important on the reef as they were in the rock pool. The sea brings in food with every tide, and the coral animals and many others filter it from the water.

At the top of the food chain are huge predatory fish called groupers, which can grow to 3m (10ft) long. Sharks and barracuda may visit the reef, to hunt the shoals of small reef fish, and take the energy out of the reef to the open sea.

The largest and best known coral reef is Australia's Great Barrier Reef, which extends along the Queensland coast.

Typical of the primary consumers of a coral reef are parrot fish, moorish idol and surgeon fish, which graze on the coral, and snails and sea urchins, which feed on algae. Secondary consumers shown here are the crab, starfish, ray, trigger fish, grouper fish and, largest of all, the shark.

Coral

Green parrot fish

Star fish

Blood-spotted crab feeds on sea urchin

Ray (feeds on burrowing snails and worms)

Sea cucumber (filter feeder)

Corals filter feed detritus

A food chain in a frozen ocean

The Antarctic Ocean supports very large numbers of animals. These can be classed into four main groups: seabirds, seals, fish and whales.

In winter, most of the higher consumers move away as the water freezes, but primary producers and consumers stay behind, because they are too small to leave. What happens to them has been discovered only recently.

During the long, dark winter, the algae are frozen into the lower surface of the ice. When the sun returns in the spring, they begin photosynthesis, even though the ocean is still frozen. They form the food of small amphipods, which are preyed on by little worms, all living in crevices in the ice.

When the ice thaws, large numbers of these plants and animals are released into the water, where larger amphipods and fish can feed on them.

Black-backed gull

Squid

Blue whale

Pollack

Among the larger amphipods are shrimps, about 3cm (1in) long, called krill. Krill live in huge swarms, weighing many tons. They are the vital bottom link of the Antarctic food chain.

Krill are eaten by seabirds, including penguins, when they return for the summer, and by fish. They are also eaten by baleen whales. These are whales such as humpbacked and blue whales. Instead of teeth, their mouths contain plates of whalebone (properly called baleen) with which they can filter the krill from the water. Baleen whales are another example of animals that take a short cut in a food chain, as does the flamingo. Although they are huge, they feed directly on the tiny krill. Krill are also the food of small squid, which are eaten by seabirds and some of the seals.

At the top of this food chain are leopard seals, which catch penguins. Another top predator is the magnificent killer whale, which hunts seals and sea lions, as well as fish.

In the icy waters of the Antarctic Ocean krill are eaten by a pollack and baby squid. The squid, in turn, are being caught by a black-backed gull. In the foreground, a fierce leopard seal hunts an adélie penguin. In the background, a huge blue whale feeds on the tiny krill.

Adélie penguin

Leopard seal

Krill

Krill
Magnified to life size

A food chain in the rain forest

This rain forest snake can swallow whole frogs and toads, which form its diet.

Rain forests grow wherever there is more than 200cm (79in) of rain per year, in places where the temperature is steady at about 25°C (77°F). Tropical rain forests are found in South America, Africa, Southeast Asia and Northeastern Australia.

A rain forest is like a green ocean, where the primary producers are at the surface. The "surface" of the rain forest is the treetop level where the leaves absorb the sun's energy and so produce food.

Animals live in the rain forest in layers, at different depths from the "surface." The food chain up among the highest branches is based on flowers, fruit and leaves. These are eaten by birds, monkeys, fruit bats and insects. The insects provide food for a few secondary consumers, such as birds, frogs and some bats, but most of the animals are herbivores.

At ground level, where there is little sunlight, the food chain depends on fruits and leaves falling from above, like detritus in the sea. Worms and many insects, including beetles and butterflies, feed on the leaf litter. They are the prey of frogs and birds. Snakes and a few small predatory mammals, such as false vampire bats, hunt the insect-eaters. Large predators of South America include ocelots, jaguars and the anaconda, a huge snake. Tigers are large predators of Asian rain forests.

Rain forests are being cut down all over the world because their trees provide valuable lumber. Many animals are dying as their natural home is disturbed, and thus the food chain is affected.

Studying food chains teaches us that all living things depend upon each other, directly or indirectly. Without the plants, which collect energy from the sun, the largest predators that gather that energy at the top of the chain could not survive.

Opposite *In the South American rain forest a howler monkey (in the background) and a squirrel monkey feed on tender leaves. A harpy eagle swoops down to catch the squirrel monkey, while a tree frog is about to snatch the leaf hopper. The blue morpho butterflies feed on nectar in flowers.*

Glossary

Algae (singular alga) Mostly single-celled plants that live in water.
Bacteria Extremely small living bodies that can cause disease or decay.
Carnivores Flesh-eaters; animals that feed mostly on other animals.
Chlorophyll A green chemical in plants that reacts with sunlight.
Colonies Groups of a single type of animal or plant living together.
Detritus The broken down remains of dead animals or plants.
Habitats The natural living places of plants or animals.
Herbivores Animals that eat plants.
Larvae (singular larva) The young that hatch from the eggs of insects.
Mammals Warm-blooded animals that are usually covered with hair or fur and feed their young on milk.
Migrate To move from one place to another at the same time every year.
Nutrients Chemicals that plants and animals need in order to grow.
Photosynthesis The chemical reaction between chlorophyll in plants and sunlight which produces food for the plant and releases oxygen into the air.
Plankton The microscopic plants and animals that drift together in oceans, lakes and rivers.
Prairies Open grassy plains of Canada and central parts of the United States.
Predators Animals that hunt and eat other animals.
Prey An animal that is hunted and killed by another animal for food.
Savannah Open grasslands found in Africa.
Scavengers Animals that feed on the remains of dead animals and plants.
Tundra The treeless Arctic region where the soil just below ground level is frozen throughout the year.

Further information

You can find out more about food chains by reading the following books:

Animal Ecology by Mark Lambert and John Williams. Franklin Watts (The Bookwright Press), 1987.
The Animal Family by Randall Jarell, illustrated by Maurice Sendak. Pantheon, 1985.
Conservation and Pollution by Laurence Santrey. Troll Associates, 1985.
Ecosystems and Food Chains by Francene Sabin. Troll Associates, 1985.
Finding Out More About Conservation by John Bentley and Bill Charlton. David and Charles, 1983.
A Look at the Environment by Margaret S. Pursell. Lerner Publications, 1976.
Once There Was a Stream by Joel Rothman. Scroll Press, 1973.
Plant Ecology by Jennifer Cochrane. Franklin Watts (The Bookwright Press), 1987.
Poisoned Land: The Problem of Hazardous Waste by Irene Kiefer. Atheneum, 1981.

Some of the wildlife documentaries shown on television will tell you more about what animals eat in the wild, and you can find out a great deal for yourself by observing wildlife in your yard or garden or a park or in the countryside.

You may want to join an organization that helps to protect wild animals and the environments they depend on. Some useful addresses are:

Audubon Naturalist Society of the Central Atlantic States
8940 Jones Mill Road
Chevy Chase, Maryland 20815
301–652–9188

Children of the Green Earth
P.O. Box 200
Langley, Washington 98260
206–321–5291

Clean Water Action Project
317 Pennsylvania Avenue
Washington, D.C. 20003
202–547–1196

The Conservation Foundation
1717 Massachusetts Avenue, N.W.
Washington, D.C. 20036
202–797–4300

Environmental Action Foundation
1525 New Hampshire Avenue, N.W.
Washington, D.C. 20036
202–745–4870

Environmental Defense Fund
257 Park Avenue South, Suite 16
New York, New York 10016
212–686–4191

Greenpeace, USA
1611 Connecticut Avenue, N.W.
Washington, D.C. 20009
202–462–1177

National Audubon Society
950 Third Avenue
New York, New York 10022
212–546–9100

National Wildlife Federation
1412 16th Street, N.W.
Washington, D.C. 20036
202–797–6800

World Watch Institute
1776 Massachusetts Avenue, N.W.
Washington, D.C. 20036
202–452–1999

World Wildlife Fund
1255 23rd Street, N.W.
Washington, D.C. 20037
202–293–4800

Picture acknowledgments

The publishers would like to thank the following for allowing their photographs to be reproduced in this book: Bruce Coleman Limited 4 (Michael Fogden); Oxford Scientific Films 6 right (G H Thompson), 11 (Peter Parks), 28 (Michael Fogden); Survival Anglia Limited 6 left and 23 (Jen and Des Bartlett); 8 and 12 (Jeff Foott).

Index

Africa, 8, 11, 22, 24, 28
Algae 7, 8, 16, 18, 23, 26
Alligators 17
Antarctic Ocean 26–7
Antelope 8, 11, 22
Arctic foxes 20
Australia 17, 25, 28

Bats 28
Bears 17, 20
Beavers 16
Beetles 4, 11, 13, 14, 17, 22, 23, 28
Birds 10, 14, 17, 20, 23, 28
  freshwater 8, 16
  garden 14
  of prey 4, 20, 23
  seabirds 19, 26, 27
  waders 19
Bison 8

Carnivores 4, 10
Caterpillars 8, 14
Consumers
  primary 8, 10, 14, 16, 20, 22, 23, 26
  secondary 10, 14, 19, 28
Coral reefs 24–5
Coypu 16
Crabs 10, 13, 19, 24

Detritus feeders 18, 19, 25,
Dragonflies 17

Elephants 8, 23
Energy 4, 6, 7, 10, 11, 12, 13, 25, 28

Filter-feeding 8, 16, 18, 19, 25, 27

Fish 10, 17, 19, 24–5, 26–7
  barracuda 25
  grouper 24
  parrot 25
  pike 17
Flamingoes 27
Flies 16, 17, 20, 22
Frogs 17, 28
Fruit 14, 28

Gardens 14
Geese 20
Grass 8, 14, 20, 22
Grasslands 7, 8, 23

Herbivores 8, 10, 11, 23, 28
Hunting dogs 13, 22
Hyenas 22

Insects 14, 20, 22, 28

Krill 27

Ladybugs 14
Lakes 16, 17, 19
Lemmings 20
Lions 11, 22

North America 16, 17
Nutrients 4

Otters 17

Penguins 27
Pests, garden 14
Photosynthesis 6, 26
Plankton 10, 25
Plants 7, 8, 11, 14, 16, 18, 20, 24, 26
Platypus 17

Predators 4, 12, 13, 14, 17, 18, 20, 22, 25, 26, 28
Producers, primary 7, 14, 16, 18, 20, 22, 24, 26, 28

Rain forests 4, 7, 28
Reindeer 20
Rhinoceros 8, 23
Rock pools 18, 19, 25

Scavengers 13, 19, 22
Sea urchins 18, 19, 24
Seals 26, 27
Sharks 12, 13, 25
Shrews 4, 14
Slugs 4, 14
Snakes 23, 28
South America 16, 28
Southeast Asia 28
Squid 27

Tadpoles 16, 17
Tigers 12, 13, 28
Trees 8, 14, 20, 28
Tundra 20

Voles 20

Water fleas 16, 17
Water snails 18, 24, 25
Whales 26–7
Wildebeest 8, 22
Wolves 20
Worms 10, 14, 16, 17, 19, 26, 28

Yards 14

Zebra 22

32